OUTBRE

The Coronavirus Pandemic

Stephen Dingus

About the Author: Steve Dingus is an educator and writer. He has taught all levels, young and old, in topics as diverse as math, science, writing, and literature. He currently teaches high school and college writing and literature. He lives near Richmond, Virginia with his wife, children, fish, and Catahoula hound dog, Lucy.

To my own children, Liam and Isabel, and my students, wherever you may be.

PUBLISHER CATALOGING-IN-PUBLICATION DATA.

Outbreak! The Coronavirus Pandemic / by Stephen Dingus – 1st ed.
ISBN 978-1-7352864-0-2 (pbk.) – ISBN 978-1-7352864-1-9 (eBook)
1. Coronaviruses Juvenile literature 2. Viruses Juvenile literature
3. Diseases Juvenile literature

First Edition

Contents

A Mysterious Illness

Late in 2019, doctors in Wuhan, China began seeing patients with symptoms like shortness of breath, high fever, and cough. Was it the flu? A bad cold? No, this was something else. In January of 2020 news media reported a 61-year-old man died from this new illness. Doctors in China soon realized a virus caused this dangerous new disease, a *coronavirus*.

Wuhan, China, where doctors first noticed the strange illness that became known as COVID-19

A few months after those first illnesses and death, the world began to take notice. In February 2020, two people in Italy—both over 70 years old—died. A few weeks later, Spain saw its first death from the new virus. In March, the virus began to spread throughout Europe. The novel coronavirus that causes COVID-19 began to take its toll on the world and change our lives.

After the first cases in Wuhan, the virus began to spread all over the world.

After infecting people in Wuhan, the virus began to spread worldwide.

What are Viruses?

A coronavirus is a type of *virus,* a very small **particle** of matter that must invade living cells to reproduce. Viruses infect bacteria, plants, and animals, including humans. But before a virus infects an **organism**, it must enter the organism's body.

Viruses transmit, or move, from one organism to another through air, mucus, or blood. A rhinovirus, the virus that causes a common cold, floats through the air after an infected person sneezes or coughs. The rabies virus passes from an infected animal's saliva. Hepatitis must transmit directly from an infected person's blood to another person's blood. No matter how a virus enters the body, once inside it must find **host** cells to infect. The virus then multiplies, spreads, and stimulates the **immune system**.

Coughing and sneezing can transmit, or pass on, a virus from one person to another.

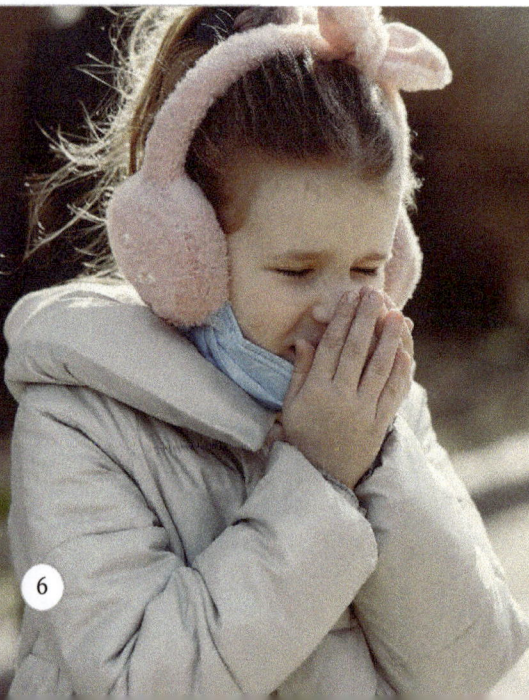

When a virus invades your body, your immune system fights back. Let's say your friend has a cold. When she coughs, water droplets carrying the rhinovirus float through the air and into your nose. The rhinovirus then infects cells lining your nose and you begin to feel ill. Your temperature rises. Your lungs generate **phlegm**. You cough. These symptoms are all part of your body's battle against the virus. Fever makes it harder for the virus to survive. Phlegm traps the virus. Your cough helps get it out. Though these symptoms eventually help get rid of the virus, they make you feel terrible.

When viruses invade our bodies, our bodies fight back by raising our body temperature, producing mucus and phlegm, and sneezing and coughing.

Our immune systems fight back when a virus invades our bodies.

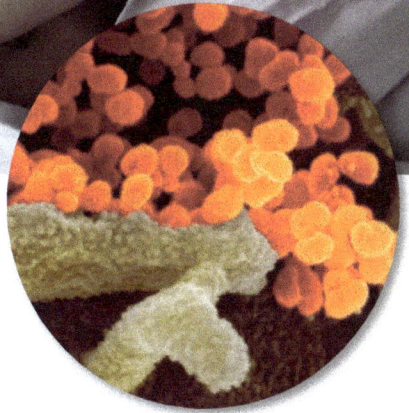

How does a virus infect the body?

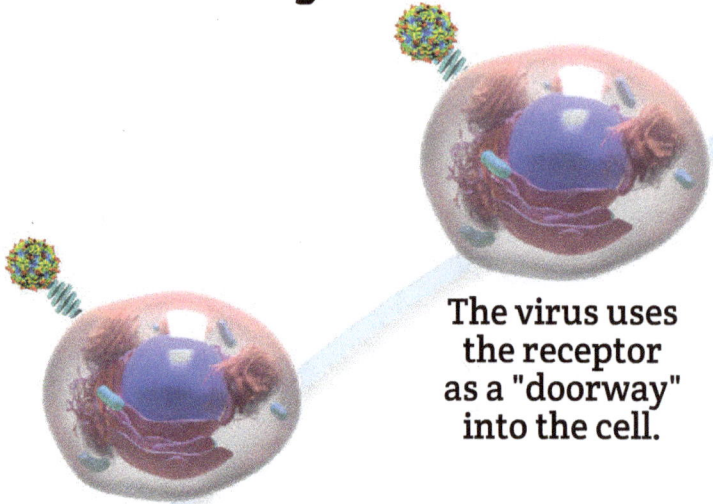

The virus uses the receptor as a "doorway" into the cell.

Once inside, the virus latches on to a receptor in a host cell. A host cell is a cell in the body the virus uses to reproduce.

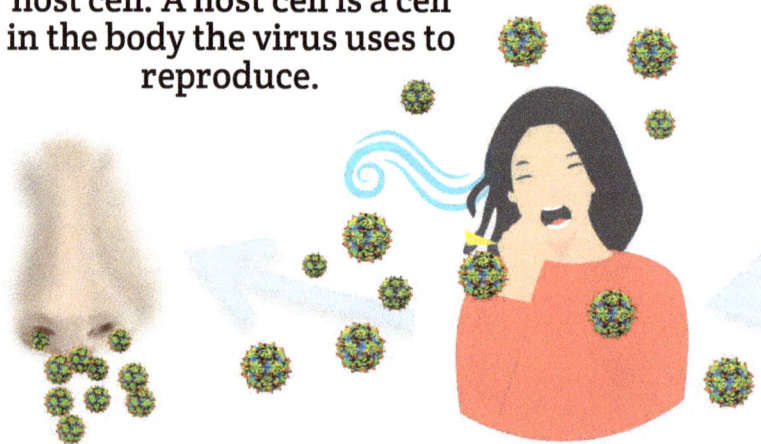

The virus must first enter the body. A rhinovirus enters the body through the nose. Other viruses might enter the body through saliva or blood.

The person coughs out the virus.

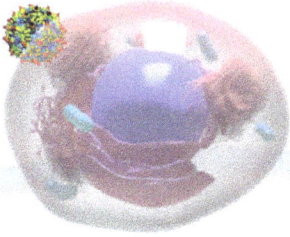

The virus
enters the cell.

The virus enters the
nucleus, the "brain" of
the cell.

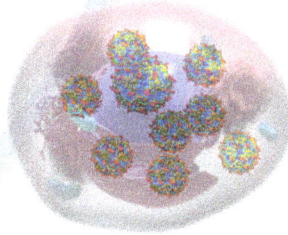

Once inside the
nucleus, the virus uses
the cell's own systems
to reproduce.

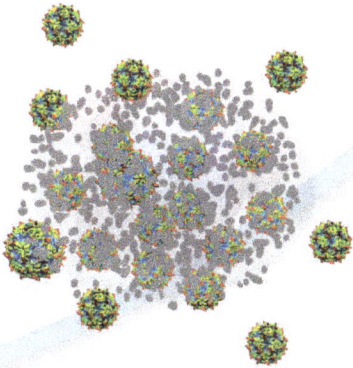

Eventually, the virus
overwhelms the cell. The
cell then bursts, releasing
the virus into the body to
infect other cells.

Viruses affect different organisms in different ways. The rabies virus, for example, attacks the **central nervous system** of mammals. The norovirus affects the **digestive system**, causing vomiting and diarrhea. Influenza infects the cells of our **respiratory system**, triggering fever, cough, and chest congestion. Viruses cause chicken pox, polio, Ebola, measles, colds, and many other illnesses. Each virus infects different parts of the body and causes different symptoms based on how our body reacts to the infection.

This is a norovirus infection seen through a scanning electron microscope. Norovirus invades cells in the human digestive system, causing nausea, vomiting, and other unpleasant symptoms.

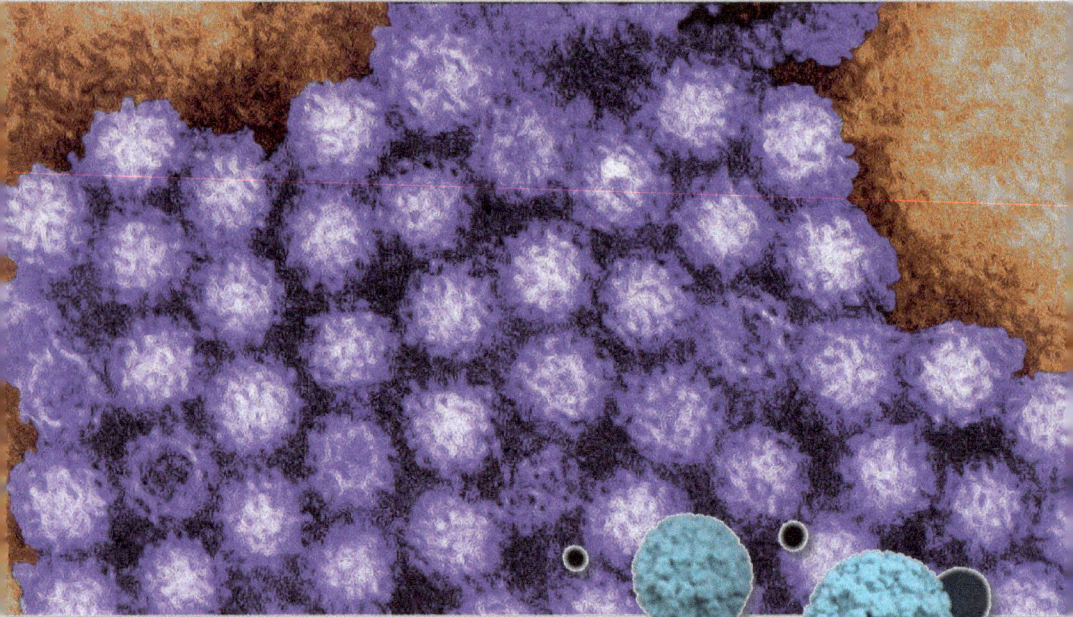

A *virion* is a single virus particle. This illustration shows norovirus virions.

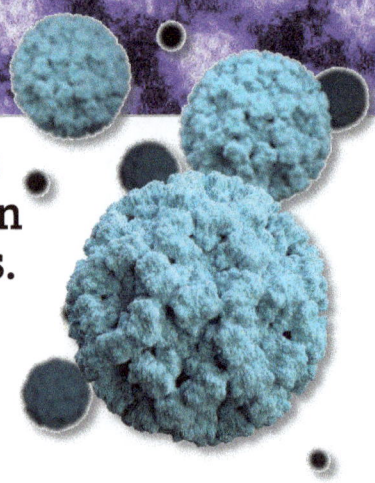

Four Common Viruses

Coronavirus

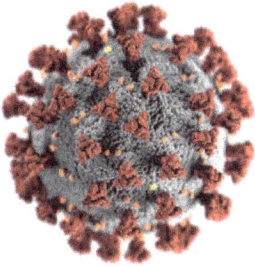

Depending on the strain, Coronaviruses can cause serious illnesses, like SARS, or mild upper respiratory illnesses, like common colds.

Influenza

The influenza virus causes the flu.

Rhinovirus

The rhinovirus causes common respiratory illnesses, like common colds.

Rabies

Rabies causes severe symptoms to the central nervous system of mammals. Luckily, rabies in humans is rare.

For years, scientists debated the exact definition of a virus. Is it a poison? A living organism? Something else? Most scientists now agree that a virus is a **molecule** in between two states—living and nonliving. Like living things, a virus can **replicate** itself. However, to replicate, a virus *must* infect the cell of a living thing. Since viruses must use another living thing to reproduce, it is impossible to classify them as living or nonliving. In some ways, they *are* alive. In other ways, they are not.

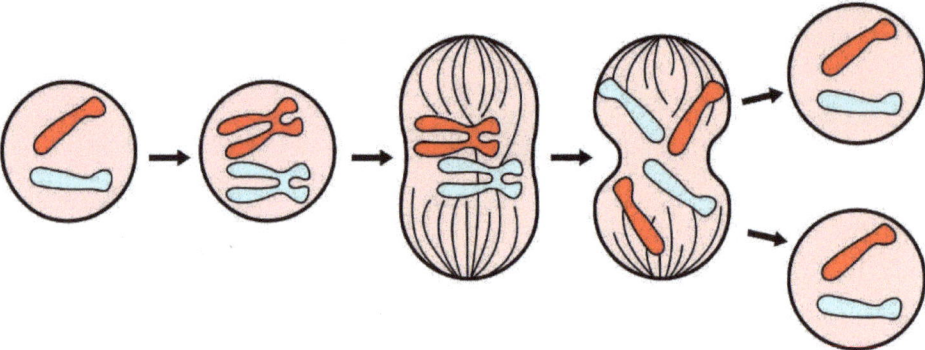

Cells of living things reproduce through cell division. Each cell has the material it needs to reproduce. Genetic material separates and eventually divides into two cells. We call this process **mitosis**.

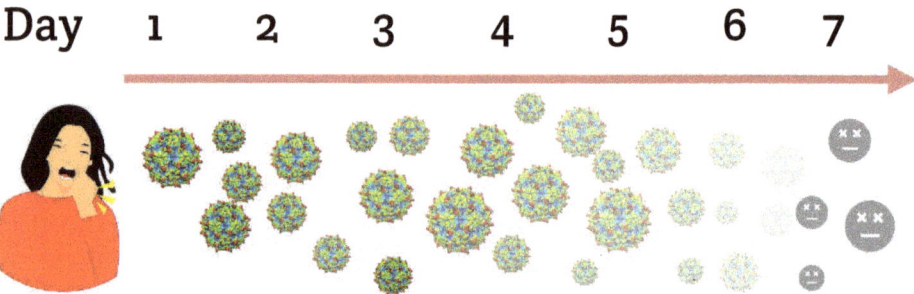

Viruses must infect cells to reproduce. Most viruses cannot survive very long outside of host cells. The rhinovirus can "live" for about seven days on surfaces outside the body. If it does not infect a host cell by then, it is no longer infectious, or able to invade a living cell.

Coronaviruses

In 2002 a man in the city of Foshan, China had a fever and dry cough. When the illness spread to his lungs, he found it difficult to breathe. Soon his wife, 22-year-old daughter, and aunt began feeling the same symptoms—a dry cough, fever, weakness, and difficulty breathing.

This new type of **pneumonia** (nuh-MOH-nya) baffled doctors. It was more severe than usual respiratory infections. They found no evidence of bacteria and it was not caused by any common virus. This was the first known case of severe acute respiratory syndrome, or SARS. Doctors soon learned that a coronavirus caused this new illness. They named it SARS-CoV, short for *severe acute respiratory syndrome coronavirus.*

A scientist at the Centers for Disease Contol and Prevention (CDC) reviews data in 2003 during the first SARS outbreak.

Though SARS-CoV was new to scientists, coronaviruses were not. Doctors discovered the first known coronavirus in 1930. For years, farmers lost large numbers of chickens to a respiratory disease called **infectious bronchitis**. Scientists discovered a previously unknown virus caused the illness—a coronavirus.

Thirty-five years later, in 1965, doctors Arthur Tyrrell and M. L. Bynoe discovered that coronaviruses cause similar illness in humans. These doctors removed sample cells lining the throat of a man with common cold symptoms, transferred these cells to a group of volunteers, and studied the results. Most of the volunteers became ill. Tyrrell, Bynoe, and other scientists soon learned a coronavirus caused these cold symptoms, similar to the virus that infects chickens. The virus particle had thorny spikes that looked like a corona, or crown. Scientists eventually named this new virus *coronavirus*.

The name coronavirus *comes from the small spikes that poke out from the virus like a corona, or crown.*

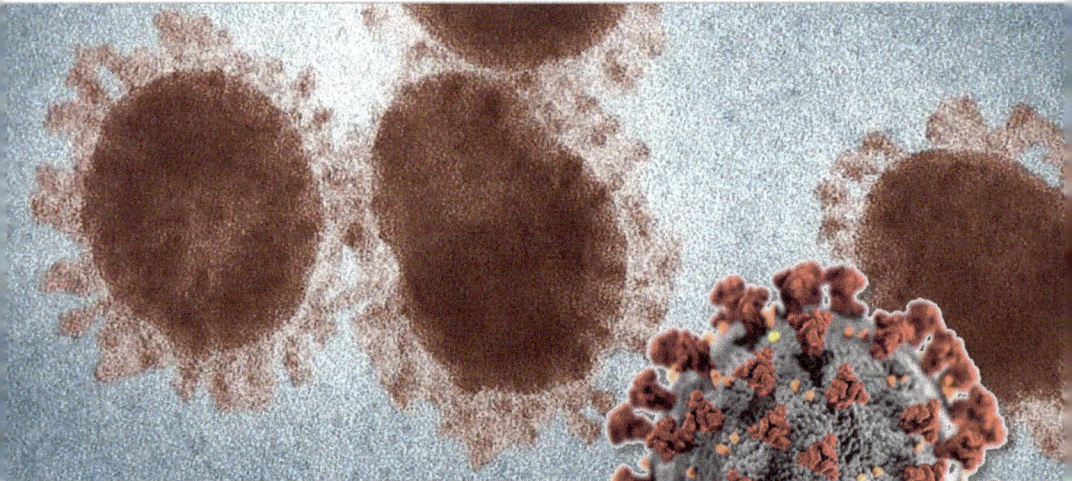

This illustration shows a coronavirus virion. You can see the thorny spikes that give the virus its name.

In 2002 a farmer in southern China became ill with a new type of pneumonia. Pneumonia is a disease that causes fluid in the lungs, making it hard to breathe.

We now know there are seven types of coronavirus that infect humans. Four of these cause mild symptoms like cough, runny nose, and aches. In addition to rhinoviruses, these four strains, or types, of coronavirus cause the common cold. But three other strains of human coronavirus infect the lungs and can make people very sick.

In 2012, ten years after the first SARS outbreak in Foshan City, a similar illness erupted in the Middle East, a region of Asia and Northern Africa just southeast of Europe. Scientists called this new illness *Middle East respiratory syndrome*, or *MERS*. Just as in China in 2002, a coronavirus caused this new illness. Scientists called this new virus *MERS-CoV*, short for *Middle East respiratory syndrome coronavirus*. Like SARS, MERS infected the lungs, causing cough, fever, and shortness of breath. The world braced for another health crisis.

Types of Coronavirus

The name *coronavirus* comes from the crown-like spikes that poke out from the virus. There are seven types of coronavirus that currently infect humans.

Coronaviruses that cause mild illness like a cold

HCoV-229E
Human coronavirus-229E

HCoV-NL63
Human coronavirus-NL63

HCoV-OC43
Human coronavirus-OC43

HCoV-HKU1
Human coronavirus-HKU1

Coronaviruses that can cause more serious illness of the lungs

MERS-CoV
Causes Middle East respiratory syndrome, or MERS

SARS-CoV
Causes severe acute respiratory syndrome, or SARS

SARS-CoV-2
Causes coronavirus disease of 2019, or COVID-19

A mysterious illness erupts in China, causing severe pneumonia in some people. Scientists discover that an unknown coronavirus causes this new illness, which they name *severe acute respiratory syndrome*, or SARS.

MERS

For the third time in less than 20 years, a new coronavirus causes severe illness in humans. The World Health Organization (WHO) names this new illness *coronavirus disease of 2019*, or COVID-19.

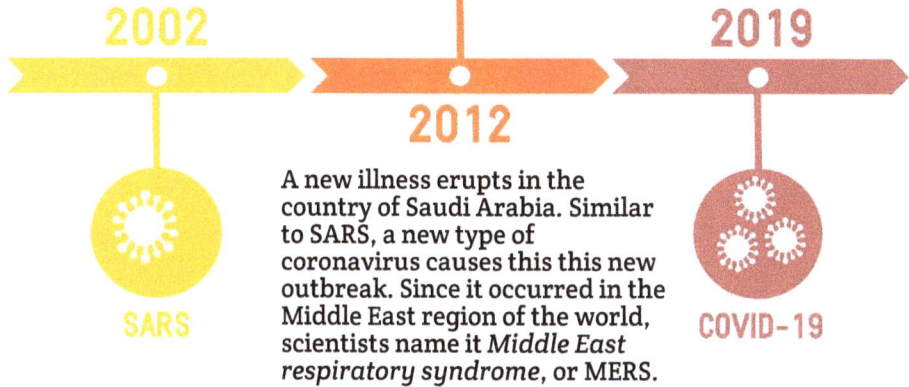

2002

2019

2012

SARS

A new illness erupts in the country of Saudi Arabia. Similar to SARS, a new type of coronavirus causes this this new outbreak. Since it occurred in the Middle East region of the world, scientists name it *Middle East respiratory syndrome*, or MERS.

COVID-19

In late 2019, it happened again. Doctors in Wuhan, China began to see the same type of symptoms they did in 2002 and 2012. But this time the virus seemed to spread from person to person easier than during those two outbreaks. Clusters of illness spread throughout the city. The government closed off Wuhan to prevent further spread. But viruses do not recognize borders. The virus soon spread outside of China and throughout the world. Just as in 2002 and 2012, scientists soon discovered a new strain of coronavirus caused the illness in Wuhan.

In February of 2020, the World Health Organization (WHO) gave this new illness its official name: COVID-19, short for coronavirus disease of 2019. The WHO named the *virus* SARS-CoV-2, short for severe acute respiratory syndrome coronavirus 2. You may have heard many names for this new virus: *coronavirus, novel coronavirus, COVID-19, COVID, SARS-2,* and others. These all refer to the new coronavirus that appeared in humans in late 2019.

What's in a Name?

You've probably heard and read a lot of new names and terms related to the COVID-19 outbreak. It is easy to get confused. Here is a summary of important terms and exactly what they mean.

Coronavirus

A type of virus that causes respiratory disease. Some strains cause minor cold symptoms. Other strains can cause serious illness like pneumonia.

COVID-19

The name the WHO gave to the illness first identified in 2019 caused by the coronavirus strain known as SARS-CoV-2.

Epidemic

A disease that spreads throughout a community or region.

MERS

Short for Middle East respiratory syndrome, the severe illness that erupted in 2012 caused by a coronavirus. The MERS-CoV virus originally infected camels.

Novel Coronavirus

Novel means new or not previously identified. A novel coronavirus is a new coronavirus scientists have not yet seen.

Pandemic

A pandemic is an epidemic that spreads to more than one country or region and can affect the entire world.

SARS

Short for severe acute respiratory syndrome, the first severe human illness caused by a coronavirus in 2002. The SARS-CoV virus originated in bats.

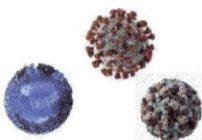

Virus

A substance in the body that infects cells and causes illness. Though the first known use of the term virus was in 1599, 1898 marks the first use of its current meaning.

Flattening the Curve

This refers to the idea that new cases of illness are no longer increasing but are staying the same. This is viewed as a flat line on a graph, rather than a steep climb.

COVID-19 changed our lives. Suddenly, local and state governments encouraged people to stay inside. Schools and colleges closed, moving classes online. Restaurants, hair salons, movie theaters, and many other businesses temporarily closed to keep people from forming large groups. When people did go out, many chose to wear masks to help protect themselves and others from spreading the virus. Many people were nervous they might contract, or catch, the virus or pass it on to loved ones.

COVID-19 changed our lives. Things we once took for granted, like going to school, suddenly stopped.

From Animals to Us

But why? Why did SARS, MERS, and COVID-19 suddenly appear in humans? Why does it cause people to get so ill when other strains of coronavirus only cause a cough and chest congestion?

Chinese scientists study data during the first SARS outbreak in 2003.

While some strains of coronavirus cause only mild symptoms, others can make people very ill.

To understand why, we must turn to one of the animal kingdom's most effective virus carriers: bats. Bats evolved special immune systems that don't overreact to viruses and other infections. Their bodies repair cell damage more efficiently than other mammals. A virus that could sicken or kill a dog or racoon or human might linger in a bat without ever causing symptoms. This makes bats a good place for viruses to develop and change. Once viruses mutate, or change, they can "jump" from a bat to a human or other animal. Scientists think this is what happened with SARS-CoV-2, the virus that causes COVID-19.

A bat hangs from a tree branch. Scientists think SARS-CoV-2, the virus that causes COVID-19, started in a bat.

A *vector* is an organism that can carry a virus and pass it on to other organisms. Bats are a vector for several viruses that can be dangerous to humans, including SARS-CoV-2.

At some point, a bat infected with the SARS-CoV-2 virus came in close contact with a person and this virus "jumped" from the bat to the person. Another theory is the virus jumped from a bat to another animal, then jumped from this **intermediary** animal to a human. Pangolins, civet cats, snakes, and turtles are all possible species for this intermediary animal, but scientists are still studying the virus's structure and genes to see exactly where it came from. Once the virus infected its first human, this person began passing the virus to other humans, starting the epidemic.

Viruses can "jump" from one species to another

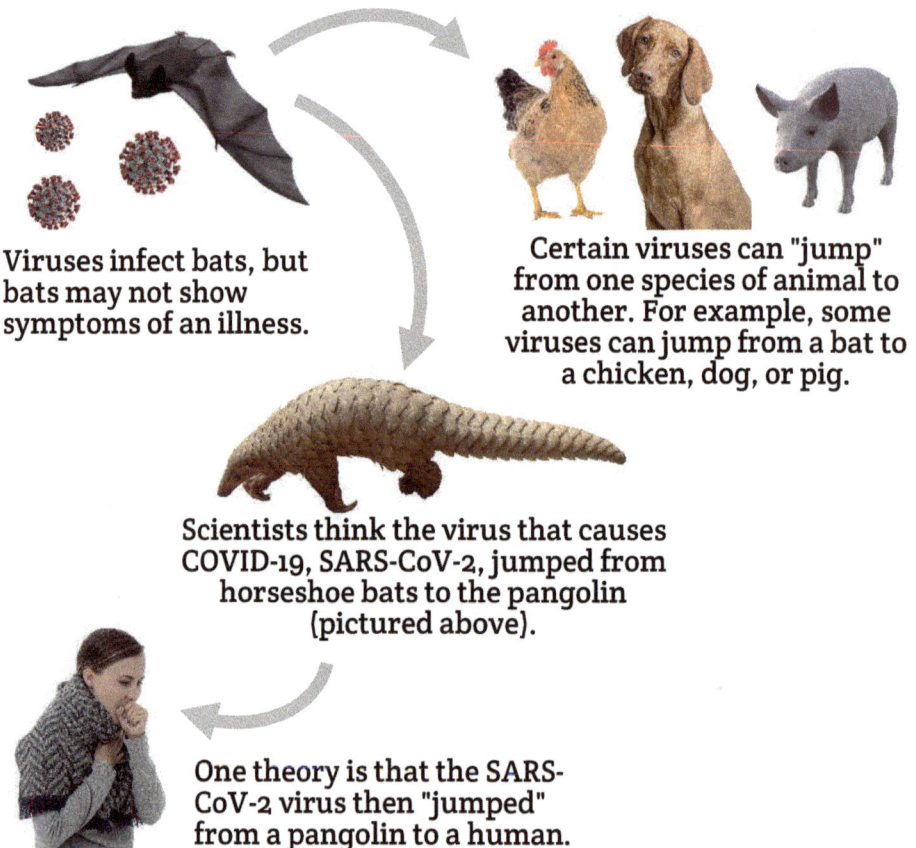

Viruses infect bats, but bats may not show symptoms of an illness.

Certain viruses can "jump" from one species of animal to another. For example, some viruses can jump from a bat to a chicken, dog, or pig.

Scientists think the virus that causes COVID-19, SARS-CoV-2, jumped from horseshoe bats to the pangolin (pictured above).

One theory is that the SARS-CoV-2 virus then "jumped" from a pangolin to a human.

No matter how it happened or which animal is responsible, the coronaviruses that cause SARS, MERS, and COVID-19 make humans ill—sometimes very ill. Since these viruses are new to humans, we have not evolved an effective *immune response*, the way our body responds to an infection. Our immune system might fight off COVID-19 as it would a mild cold. But since this is a *novel* coronavirus—a virus our body has not yet seen—our immune system could also go haywire. It might, for example, fill lungs with fluid to fight the virus, causing pneumonia. Doctors and nurses who work with COVID-19 patients also face danger. They inhale so much of the virus that it simply overwhelms the immune system's ability to fight it off.

Since SARS-CoV-2 is a *novel* coronavirus—a virus our body has never seen—our immune system might not know how to react.

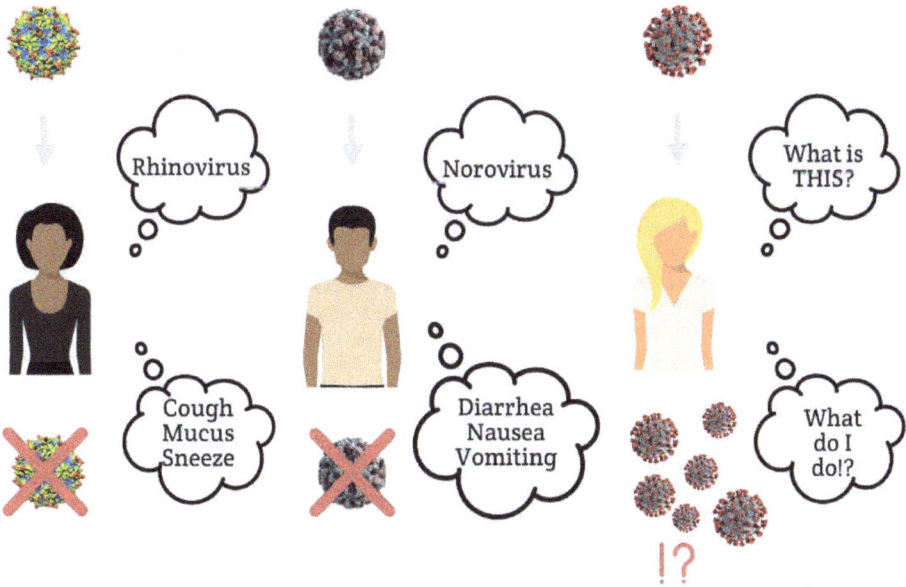

Rhinovirus

Cough
Mucus
Sneeze

Norovirus

Diarrhea
Nausea
Vomiting

What is THIS?

What do I do!?

!?

Staying Safe

So, how can you stay safe? First, don't panic, but be smart. COVID-19 usually infects older adults. It can infect children, but less than 2% of all people with COVID-19 are under 18. Children and adults should take precautions. If you take some basic steps, you can protect yourself and your family from catching many illnesses, including COVID-19. *See the infographic on page 25 for ways to stay healthy and protect yourself from viruses.*

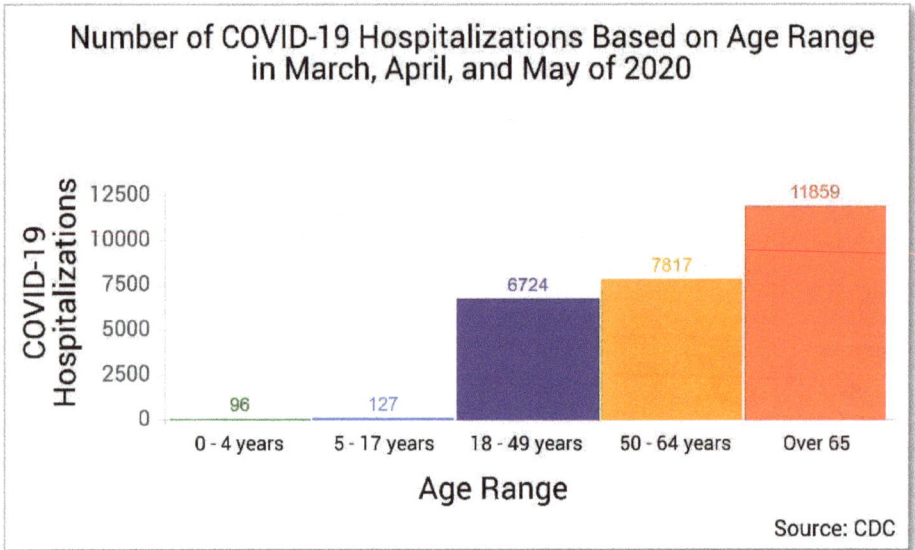

This graph shows the number of people entering the hospital over a three month span due to COVID-19 by age range. Less than 1% are children under 18.

But even if children do not get symptoms of COVID-19, they can still transmit the virus to friends and family. SARS-CoV-2, the virus that causes COVID-19, can be dangerous to people who are elderly or already ill with another disease. Because of this, children, like adults, should be careful they do not spread the virus.

Stay Healthy and Safe

Here are some basic steps you can take to keep yourself and loved ones safe from any virus, including the virus that causes COVID-19.

Wash your hands

Regularly wash your hands with soap and water for at least 20 seconds, especially after going out in public. Sing *Happy Birthday* to yourself to measure 20 seconds.

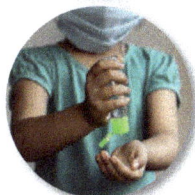

Use hand sanitizer

Use alcohol-based hand sanitizer after washing your hands or when you are unable to wash your hands. Hand sanitizer kills the virus.

Don't touch your face

Avoid touching your eyes, nose, and mouth. Hands can transfer the virus to your eyes, nose, or mouth and infect your body.

Tell an adult if you're sick

If you have a cough, fever, or trouble breathing, tell a parent, guardian, or trusted adult right away.

If you're sick, stay home

If you do have a cough or fever, stay home, even if your symptoms are mild. You might infect other people with the virus that causes COVID-19. Also, sneeze or cough into a tissue or your elbow to avoid spreading the virus.

Stay informed

Ask your parents or a trusted adult to keep you updated on news about COVID-19 or read or watch a kid-friendly news source.

Masks are another important way we can help stop the spread of COVID-19. Masks can prevent transmission of the virus by stopping droplets of saliva and **mucus** from entering the air when we breathe or cough. Though someone might not have symptoms, they can still carry the virus. So, wearing a mask keeps people with mild symptoms or no symptoms from spreading SARS-CoV-2. Wearing a mask also filters air we breathe *in*. It helps keep the virus *out* of our respiratory system, preventing infection.

Masks help prevent saliva droplets that may contain a virus from entering the air.

Do masks really work? Scientists conducted many studies to answer this question. Scientists in South Korea, for example, had four COVID-19 patients cough into a Petri dish—a shallow, round dish used in science experiments. The patients coughed into a dish without masks, through surgical masks, and through cloth masks. The scientists then measured the number of SARS-CoV-2 molecules in each Petri dish. They found that neither mask prevented the spread of the virus.

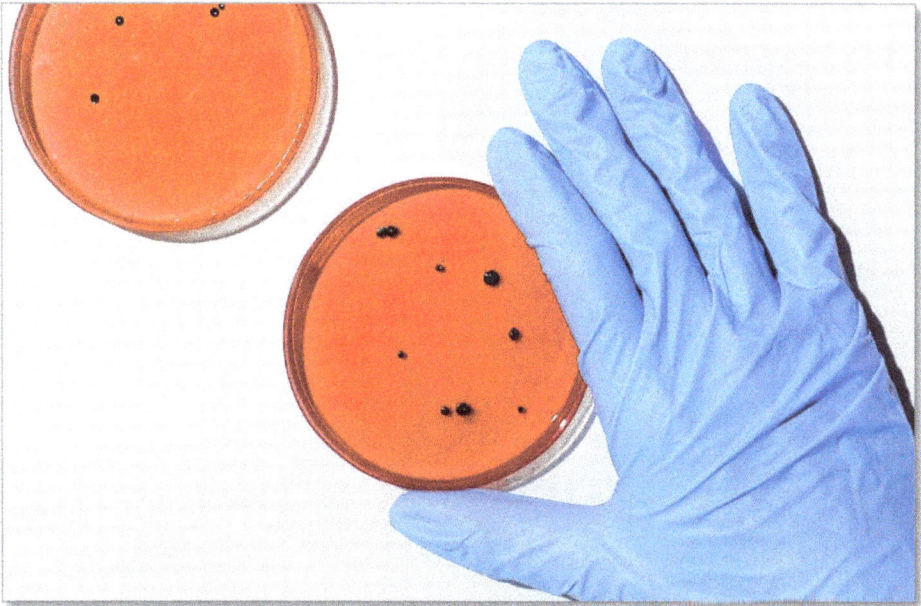

*Though some studies conclude that masks have no effect on preventing the spread of COVID-19, many other studies conclude that masks **do** help limit the spread of COVID-19.*

Other researchers, however, saw problems with this study. First, patients were only 20 centimeters from the dish. People are rarely this close to each other. Second, the study only used four patients. Most studies use more participants. Finally, the study did not measure if masks shortened the distance a virus particle travels. This is important data since it tells us how close we can be to others without danger of infection when wearing a mask.

One study in the journal *Nature Medicine* examined whether masks prevented the transmission of influenza, rhinovirus, and coronavirus in 246 patients. Scientists collected virus samples in the nose and throat of participants, then measured the amount of virus particles these participants breathed out with and without masks. When they gathered all the data, these scientists concluded that masks "could prevent transmission of human coronaviruses."

Another study in *Proceedings of the National Academy of Sciences* examined data from infected countries around the world. Scientists found that infection rates dropped when countries required their citizens to wear masks in public. Because of these studies and many others like them, the Centers for Disease Control and Prevention (CDC) recommends that everyone wear a mask when inside public places, like stores, gyms, and restaurants during a pandemic.

According to one study, masks "could prevent transmission of human coronaviruses."

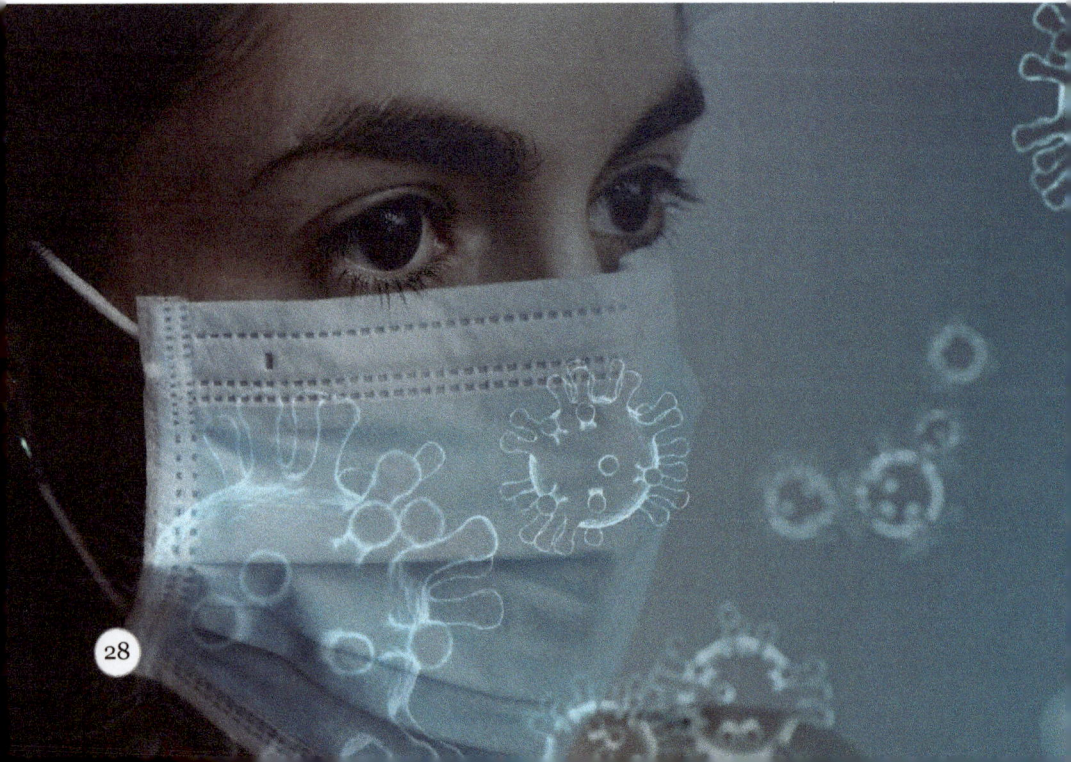

Social distancing is another tool in the battle against the transmission of coronavirus. People closer than six feet for a prolonged period have a much greater chance of contracting, or becoming infected with, the virus. Those who stay at least six feet away from other people make it harder for the virus to travel from person to person. That is why the CDC recommends people stay at least six feet away from other people, a practice called *social distancing*.

When people are closer than 6 feet, the virus can move from person to person.

3 feet   3 feet

When people are at least 6 feet apart, the virus falls before it can spread to the other person.

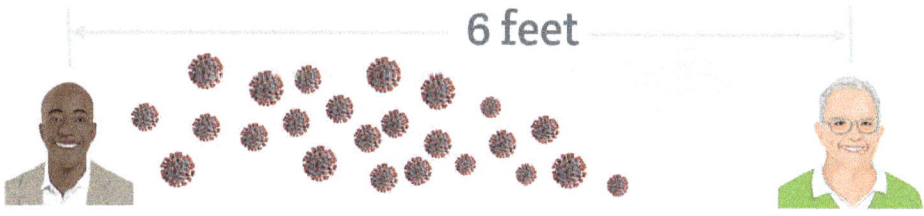

6 feet

Of course, people can't *always* stay six feet from everyone else. You must use your own judgement about masks and social distancing and decide what actions you and your family can take to keep each other safe. Talk to parents, teachers, family members, and friends about how you will be able to visit and interact safely during any pandemic.

The Spanish flu, Ebola, SARS—these illnesses affect millions of people. Many people die. Many more fall ill. But humanity persists. We use research, knowledge, and analysis to find solutions. Two devastating viral diseases, polio and smallpox, infected millions of people during the 20th century. Through hard work and persistence, scientists developed vaccines for both diseases, saving millions of lives per year. Right now, scientists of all kinds are working to develop, test, and manufacture vaccines for all types of viruses, from rabies to flu to tetanus to coronavirus... and many, many others.

COVID-19 changed our lives. Things we took for granted, things like parties and school, suddenly stopped. As we move onward into the 21st century, coronavirus is another problem we must overcome. Early in the pandemic, we entered **quarantine**. When we *were* with people, we kept a safe distance of six feet. We wore masks to prevent spreading the virus. Eventually, coronavirus will pass. But we must continue to learn, to grow, and to problem solve. The solutions to humanity's problems start with *you*. How will you change the world?

Glossary

bacteria (*bak-TIR-ee-uh*) Small, simple one-celled organisms. Some types of bacteria make people ill.

central nervous system The system of the body containing the brain and spinal cord, and that controls functions of the body including thinking and our senses

coronavirus (*kuh-RO-nuh-vye-ruhs*) A type of virus with spikes that resemble a corona, or crown. Coronavirus causes illness in many organisms.

digestive system (*di-jest-iv sis-tuhm*) The system of the body containing the tongue, throat, stomach, and other organs that breaks down food so the body can absorb nutrients

host An animal or plant—or the cell of an animal, plant, or bacteria—in which a virus or parasite lives

immune system (*i-MYOON sis-tuhm*) The system of the body that fights off disease with special organs, cells, and molecules known as *antibodies*

infectious bronchitis (*in-fek-shus brahng-KI-tuhs*) An illness in the upper respiratory system that can pass from one organism to another

intermediary (*in-ter-MEE-dee-air-ee*) Something between two things. An organism infected with a virus from one species of organism that passes it to another species of organism.

molecule (*MAHL-uh-kyool*) The smallest unit of matter that has characteristics that separate it from other matter

mucus (*MYOO-kuhs*) Fluid that lubricates and protects the mouth, throat, lungs, and other parts of the body

organism (*OR-guh-ni-suhm*) A living thing with a system of organs (body parts with separate functions) to help support life

particle (*PAR-tik-uhl*) A very small piece of matter

phlegm (*flem*) Thick mucus coughed up when a person is sick

pneumonia (*nuh-MOH-nya*) Fluid in the lungs, usually caused by a virus or bacteria, that makes it hard to breathe

quarantine (*KWOR-uhn-teen*) A period of time when a person or people with a disease must stay away from other people

replicate (*REH-pluh-kat*) To make exact copies. Cells and viruses make copies of *themselves*.

reproduce (*ree-pruh-DOOS*) To make new organisms of the same kind

respiratory system (*REH-spur-uh-tor-ee sis-tehm*) The parts of the body that allow it to breathe, or take in oxygen for the body use

stimulate (*sti-myoo-lat*) To make active or more active; to energize

Index